KILLER NATURE

TINY BUT DEADLY CRITTERS

by Charles C. Hofer

CAPSTONE PRESS
a capstone imprint

Published by Capstone Press, an imprint of Capstone.
1710 Roe Crest Drive
North Mankato, Minnesota 56003
www.capstonepub.com

Library of Congress Cataloging-in-Publication Data
Names: Hofer, Charles, author.
Title: Tiny but deadly critters / by Charles C. Hofer.
Description: North Mankato, Minnesota : Capstone Press, [2022] | Series: Killer nature | Includes bibliographical references and index. | Audience: Ages 8–11 | Audience: Grades 4–6
Summary: "The world is crawling with all sorts of bugs and other little critters. But don't be fooled by their small size. Many of these creatures are quite dangerous. Some are armed with powerful poison that can kill a person. Others carry and spread deadly diseases. Brace yourself to discover some of the world's tiniest killers"— Provided by publisher.
Identifiers: LCCN 2021004202 (print) | LCCN 2021004203 (ebook) | ISBN 9781663906311 (hardcover) | ISBN 9781663906281 (pdf) | ISBN 9781663906304 (kindle edition)
Subjects: LCSH: Poisonous animals-Juvenile literature. | Insects—Juvenile literature.
Classification: LCC QL100 .H64 2022 (print) | LCC QL100 (ebook) | DDC 591.6/5—dc23
LC record available at https://lccn.loc.gov/2021004202
LC ebook record available at https://lccn.loc.gov/2021004203

Editorial Credits
Editor: Aaron Sautter; Designer: Heidi Thompson; Media Researcher: Kelly Garvin; Production Specialist: Katy LaVigne

Image Credits
Newscom: Martin Harvey/NHPA/Photoshot, 19, Paulo de Oliveira/NHPA/Avalon.red, 15; Shutterstock: akiyoko, 21, Christian Vinces, 23, Ernie Cooper, 9, Gerald Robert Fischer, 17, IrinaK, 8, Jaco Visser, 5, 25, Jay Ondreicka, cover, 7, khlungcenter, 29, Laura Ekleberry, 1, 11, Mikhail Gnatkovskiy, 4, Pavel Krasensky, 27, unterwegs, 13

Artistic elements: Shutterstock/Fer Gregory

All internet sites appearing in back matter were available and accurate when this book was sent to press.

Printed in the United States 5042

TABLE OF CONTENTS

Words in **bold** text are included in the glossary.

TINY TERRORS

Not all deadly animals are big terrors. They can be tiny too. Some can sting. Some can bite. Some can make you very sick.

These tiny terrors can swim, and they can fly. They can crawl, and they can slither. They're found everywhere around the world. They live in oceans and deserts and forests. Some may even live near you!

Some mini killers are armed with powerful and deadly **poisons**. Some little terrors spread diseases that kill lots of people each year. And still others are dangerous bullies.

They may seem small and harmless, but be careful. Some tiny critters can be real killers!

BLACK WIDOW SPIDER

Black widow spiders are found throughout North America. They like to hide in dark areas like basements or sheds. The female black widow is easy to recognize. It has a red "hour glass" shape on its **abdomen**. But don't get too close!

These spiders have powerful poison called **venom**. The spider delivers it through a bite. It can be deadly to humans, especially to small children. A black widow's venom is 15 times stronger than a rattlesnake's.

Like all spiders, the black widow is a **predator**. It feeds on insects that it catches in its web. But that's not all. Sometimes, female black widows will eat the males!

FACT

Animals like spiders and snakes usually use venom to catch **prey**. This powerful poison is delivered through a sting or bite.

FEATURE FILE

Common Name: Black widow spider

Habitat: Quiet, dark places like basements, cupboards, or wood piles

Range: All over the world, except Antarctica

Size: Length up to 1.5 inches (3.8 centimeters)

Deadly Feature: Venomous bite

BARK SCORPION

Scorpions are closely related to spiders. They also have eight legs and an **exoskeleton**. This hard shell helps protect them. Most scorpions also have big claws.

All scorpions have stingers located at the end of their tails. The stingers are loaded with deadly venom. This helps them catch prey to eat. They also use their stingers to scare off any threats.

The tiny bark scorpion is common in the Desert Southwest. It likes to hide in dark places, such as piles of wood or leaves. Most scorpion stings are painful to people, but not deadly. However, the bark scorpion's powerful venom can kill young children or older people. It's the only deadly scorpion in the United States.

GLOW-IN-THE-DARK SKIN

Scorpions can glow in the dark! Their exoskeleton glows brightly when under a special light. Scientists aren't sure why scorpions glow in the dark. Some experts think this ability might help scorpions communicate with each other.

FEATURE FILE

Common Name: Bark scorpion

Habitat: Deserts

Range: Southwest United States, Northern Mexico

Size: Length between 1 and 3 inches (2.5 and 7.6 cm)

Deadly Feature: Venomous sting

GIANT DESERT CENTIPEDE

The desert is full of creepy critters. The giant desert centipede might be the creepiest. It looks like a tiny alien crawling across the desert floor. A young desert centipede has 8 to 10 pairs of legs. It adds legs as it grows. An adult giant desert centipede can have up to 23 pairs of legs!

The giant desert centipede is **nocturnal**. It hunts and eats insects mainly at night. It will also eat small mammals and reptiles. The centipede uses a special pair of legs to pinch its prey. Its pinch injects powerful venom into the prey. It isn't deadly to humans. But it does hurt! The desert centipede has one of the most painful pinches or "bites" in the world.

FEATURE FILE

Common Name: Giant desert centipede

Habitat: Deserts

Range: Southwest United States, Missouri, and Northern Mexico

Size: Length up to 8 inches (20.3 cm)

Deadly Feature: Painful venomous pinch

FLOWER SEA URCHIN

The ocean is home to many strange creatures. One ocean oddball is the sea urchin. It looks like a tiny ball covered in sharp spines. The sea urchin uses its spines for protection from predators. Just one poke can be very painful.

Some sea urchins can be deadly. The flower sea urchin has powerful venom. It can even kill a person! Most sea urchins deliver venom through their spines. But not the flower sea urchin. It has flowerlike parts that grow between its spines. These "flowers" are what deliver the deadly venom.

FACT

All sea urchins have dozens of tiny tubelike feet. They work like little suction cups. This allows urchins to stick to hard surfaces like rocks or coral.

FEATURE FILE

Common Name:	Flower sea urchin
Habitat:	Coral reefs, rocky areas, seagrass beds
Range:	Southeast Atlantic Ocean, Indo-West Pacific Ocean
Size:	Width up to 5.9 inches (15 cm)
Deadly Features:	Powerful venom; painful spines

GEOGRAPHY CONE SNAIL

The geography cone snail lives in the warm waters of the Pacific Ocean. This tiny snail has the most powerful venom in the world. Just one sting from a large cone snail can kill a human within hours.

The cone snail is a type of **mollusk**. Mollusks have soft, squishy bodies with no bones. Mollusks instead have a hard shell that protects them from danger.

The cone snail hunts and eats fish. It has sharp teeth called radulae. The radulae delivers the snail's powerful venom. When a fish swims by, the snail shoots its radulae like an arrow to sting its prey. The powerful venom paralyzes the fish so the snail can eat it.

FEATURE FILE

Common Name:	Geography cone snail
Habitat:	Coral reefs
Range:	South Pacific Ocean
Size:	Length up to 6 inches (15 cm)
Deadly Feature:	Venomous radulae

BEARDED FIREWORM

Bearded fireworms look like tiny, hairy centipedes from the forest. But these creepy crawlies live in warm ocean waters. Fireworms are mostly active at night. They slither around coral reefs looking for food. They like to eat small snails, clams, and other tiny ocean animals.

Bearded fireworms have bright coloring. The colors tell enemies, “Back off! I’m dangerous!” Fireworms are also covered in tiny bristles. They look soft, but you don’t want to pet them. The colorful hairs are sharp and filled with a painful venom.

The fireworm’s venom is used for defense against threats. The venom-filled hairs break off and stick into a predator’s skin. The pain can last for hours.

FEATURE FILE

Common Name:	Bearded fireworm
Habitat:	Coral reefs, rocky areas
Range:	Caribbean Sea, Mediterranean Sea, Atlantic Ocean
Size:	Length up to 13.8 inches (35 cm)
Deadly Feature:	Venomous bristles

AFRICANIZED HONEY BEES

Honey bees are very helpful insects. They help **pollinate** plants, which helps farmers grow food. This helps keep the environment healthy. Female bees have stingers loaded with venom. They use their stings for defense against threats. Bee stings can be painful. But for most people, they're usually not deadly.

However, African bees are different. These bees were brought to the United States in the 1950s. African bees can be very aggressive. They are sometimes called "killer bees."

Like all bees, African honey bees are good at working together. When someone threatens their hive, the African bees will swarm. Hundreds of bees will attack and sting at the same time. The swarm's attack can be deadly, even to humans!

FACT

Some people are **allergic** to bee stings. They can die from a single sting. It's important to always be careful around bees. But remember that they are just trying to defend themselves.

FEATURE FILE

Common Name:	Africanized honey bee
Habitat:	Everywhere in warm regions
Range:	Central America, South America, Southern United States
Size:	Length up to 0.75 inch (1.9 cm)
Deadly Features:	Venomous sting; aggressive swarms

ASIAN GIANT HORNET

Female Asian giant hornets have stingers filled with powerful venom. People say the hornet's sting feels like a hot nail piercing their skin. These huge hornets are also thought to spray venom at enemies.

The Asian giant hornet is the world's largest hornet. Along with stingers, they are also aggressive. Asian giant hornets are known to attack and kill entire hives of honey bees. No wonder people often call them "murder" hornets!

Asian giant hornets are native to Asia. The first one in the United States was found in 2019. The hornets are an invasive species. These are plants or animals that come to a new area from a different place. They are often harmful to local plants and animals.

INVASION!

An invasive species can come in any shape or size. It could be a pet that escaped into the wild. It could be a plant that grows out of control. Invasive species can also spread harmful diseases.

FEATURE FILE

Common Name: Asian giant hornet

Habitat: Forests

Range: Asia; new to North America

Size: Length up to 2 inches (5 cm)

Deadly Features: Powerful venom; aggressive nature

BULLET ANT

Bullet ants live in rain forests in South America. These insects are small. But for ants, they're huge. An adult bullet ant can grow more than 1-inch (2.5-cm) long. Bullet ants are predators. They usually eat other small insects. They will also feed on dead plants and animals. This helps keep the forest clean and healthy.

Bullet ants live in a large group called a **colony**. One colony can have up to 10,000 ants. Like all ants, they are very good at working together. When threatened, bullet ants will swarm over and attack an enemy.

Bullet ants get their name from their incredibly powerful venom. They deliver the most painful sting in the world. It usually won't kill a person. But it feels like getting shot by a bullet. The horrible pain can last up to 24 hours!

FEATURE FILE

Common Name:	Bullet ant
Habitat:	Rain forests
Range:	South America
Size:	Length up to 1.5 inches (3.8 cm)
Deadly Feature:	Extremely painful sting

TSETSE FLY

Not all tiny killers have deadly stings. Some spread deadly diseases. One such super spreader is the tsetse (TET-see) fly. This tiny fly is common in many parts of Africa.

Tsetse flies feed on the blood of mammals, like cows or humans. When a fly lands on a victim, it uses its **proboscis**. This is part of the fly's mouth. It's long and sharp like a needle. It pierces the animal's skin, causing blood to flow. Ouch! The fly then sucks up the blood through its mouth parts like a straw.

But the tsetse fly's bite isn't just painful. It can be deadly. The fly spreads a disease called sleeping sickness through its bite. The disease causes horrible headaches and other body pains. Victims also have trouble sleeping normally. Unless victims are treated, they will eventually die.

FEATURE FILE

Common Name: Tsetse fly

Habitat: Deserts, dry areas

Range: Africa

Size: Length up to 0.7 inch (1.7 cm)

Deadly Feature: Spreads disease

FLEAS

Fleas are very tiny insects. There are more than 2,500 types of fleas around the world. Fleas often live on pets like cats or dogs. They are **parasites**. These pests usually hide in an animal's fur or feathers. Fleas bite through the skin and then feed on the animal's blood.

Fleas mostly bite and cause a lot of itching. But sometimes they're deadly. Fleas can help spread **plague**. This horrible disease can kill many people. During the 1300s, fleas helped spread a plague in Europe and Asia. The disease became known as the "Black Death." About 25 million people died from the Black Death.

FACT

The plague still occurs in the United States today. Some animals, such as prairie dogs, can carry the plague.

FEATURE FILE

Common Name: Flea

Habitat: Everywhere

Range: Everywhere mammals and birds live

Size: Length up to 0.13 inch (0.33 cm)

Deadly Feature: Spreads disease

MOSQUITOES

They buzz in your ear. They can ruin a day at the park. Their bite causes a terrible itch. Mosquitoes are no fun. But in some places, they're more than just annoying. Mosquitoes can be deadly. In fact, the mosquito is the deadliest animal in the world.

Mosquitoes bite animals and people to feed on their blood. The pests often spread deadly diseases through their bite. Yellow fever and West Nile virus are two horrible diseases spread by mosquitoes. The diseases are mostly found in places like Africa and South America. Diseases spread by mosquitoes kill about 1 million people every year.

Be careful out there! Although some critters are tiny, they can still be dangerous or even deadly. It's good to know any dangerous creatures that might live near you.

FEATURE FILE

Common Name: Mosquito

Habitat: Everywhere; prefers wet, warm areas

Range: Worldwide

Size: Length up to 1 inch (2.5 cm)

Deadly Feature: Spreads disease

GLOSSARY

abdomen (AB-duh-muhn)—the lower part of an insect's or spider's body

allergic (uh-LUHR-jik)—a condition in which a person's body has a strong reaction to something, such as pollen or insect stings

colony (KAH-luh-nee)—a large group of insects that live together

exoskeleton (ek-soh-SKE-luh-tuhn)—the hard outer shell of an insect that helps protect it

mollusk (MAH-lusk)—a soft-bodied creature that usually has a shell

nocturnal (nok-TUR-nuhl)—active at night and resting during the day

parasite (PAIR-uh-site)—an animal or plant that lives on or inside another and causes harm

plague (PLAYG)—a serious disease that spreads quickly and often causes death

poison (POI-zuhn)—a substance that can cause illness or death

pollinate (POL-uh-nayt)—to transfer pollen from one flower to another; pollen allows flowers to make seeds

predator (PRED-uh-tur)—an animal that hunts other animals for food

prey (PRAY)—an animal that is hunted and eaten by another animal

proboscis (proh-BOS-is)—a long, tube-shaped mouthpart

venom (VEN-uhm)—a poisonous liquid produced by some animals

READ MORE

Lawrence, Ellen. *Poison Slimers: Poison Dart Frogs, Sea Cucumbers, and More*. New York: Bearport Publishing, 2019.

Levy, Janey. *Lethal Insects*. New York: Gareth Stevens Publishing, 2020.

Turner, Matt. *Flying Creepy Crawlers*. Minneapolis: Hungry Tomato, 2017.

INTERNET SITES

BugFinder: Insect Search and Identification
insectidentification.org/bugfinder-start.php

National Geographic Kids: Wacky Weekend: Deadly Creautures
kids.nationalgeographic.com/wacky-weekend/article/deadly-creatures

Top 10 Most Poisonous Bugs in the World
youtube.com/watch?v=-YOrs6Q_mSk

INDEX